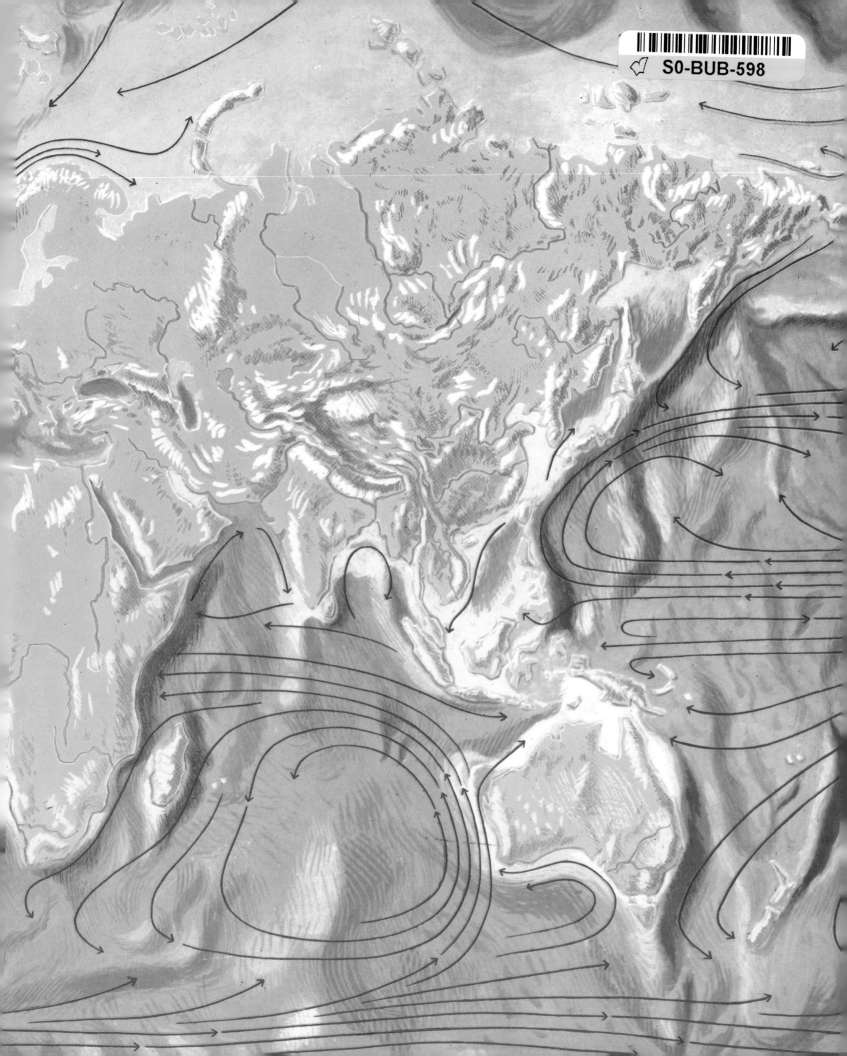

HEIN WENZEL · THE ETERNAL SEA

THE OCEAN AND THE COAST   15

THE BIG DEEP   117

STORMY SEAS   125

TRADE WINDS AND MONSOONS   138

MAN AND THE SEA   153

THE ETERNAL SEA

A PICTORIAL ANTHOLOGY by HEIN WENZEL

Translated by WALTER KAUFMANN

ABELARD-SCHUMAN
London · New York · Toronto

MY GRIEF ON THE SEA

My grief on the sea,
How the waves of it roll!
For they heave between me
And the love of my soul!

Abandon'd, forsaken,
To grief and to care,
Will the sea ever waken
Relief from despair?

My grief and my trouble!
Would he and I were,
In the province of Leinster,
Or County of Clare!

Were I and my darling –
O heart-bitter wound! –
On board of a ship
For America bound.

On a green bed of rushes
All last night I lay,
And I flung it abroad
With the heat of the day.

And my Love came behind me,
He came from the South;
His breast to my bosom,
His mouth to my mouth.

Douglas Hyde

E. Nolde: Yachts in the Yellow Sea

FOREWORD

Ingenious nautical devices have taken enough of the hazards out of seafaring to make the passage of modern ships reasonably predictable. For all that, now as ever, we need men aboard, not seagoing transport workers, we need men that won't keel over in tropical heat, won't whimper in corners when the icy blades of winter cut without mercy, men prepared to brave what dangers of the sea remain.

There are compensations.

Truly, the sea has its magic – a sunrise on the skyline and the hue of the waves changing in the light of the rising sun, a shoal of fish moving swiftly in the cooler currents, dolphins maybe, free in the vast sea and swifter even than the ship, whales in the distance spouting water like fountains, or the flight of an albatross high above the masts in the dome of the sky; and yes, there is romance in the emergence of an unknown shore: coral reefs of fabulous colours before a wide yellow beach fringed with palms, white cliffs stark and high over the sea, a clouded mountain range far beyond the shoreline. And there is romance in the promise of the sight of a harbour, in the promise of a short leave ashore: laughter, wine and whisky drunk in fellowship, and the love of women, young, beautiful and generous, of which a seaman only dreams since all too often the dream surpasses reality . . .

It takes time to know the sea, years, and those who keep measuring the brief moments of romance against the punishment of mind and body that the sea imposes, soon gravitate toward a sullen hatred which is akin to the hatred of an alien land where none speak your tongue and all customs are strange. Accursed be the breakers pounding the deck, surging in fierce succession from stem to stern, accursed the torrential rains tumbling from an ominous sky, the storms that howl through the riggings, the valleys into which the ship plunges, not once, but again and again and again, not for hours, but for days, sometimes for a week and more, day and night. Accursed be those moments of deadly fear when the ship groans and shudders as the screw races wildly in the hollow valleys of the waves, the engines thumping, the anchor chains thundering with a fearful sound in the hawse pipes, and now the cargo shifts, now the ship itself is listing, and still she plunges into the valleys of the waves. Have you lived through that, ever? If you have, accursed be for all time that calm before the storm when not a ripple disturbs the mirror of the sea, no wind blows and no clouds move, the tropical heat oppressive even at dawn when the red ball of an alarming sun rises to the zenith.

Accursed be all that, and more:
What of the icefloes in winter, huge and menacing, and the ice forming on deck and the weight of ice pressing the ship below the waterline. And what of a treacherous fog in the last days of summer, say, when it is not yet expected, a fog so thick you cannot see the bow from the bridge, nor the top of the masts, nor even any part of them, the fog-horn eerie in the night and the ship's speed throttled to barely perceptible movement, and the sound of the invisible sea subdued in the fog!
Ingenious nautical devices – radar, automatic steering, automatic alarm – Save Our Souls! SOS!
How many facets has a gem? The sea has more.
How many colours has the rainbow?
The sea's palette is not poorer.
How many moods has man, how many contradictory emotions propel him into action and toward his destiny?
The sea knows all the moods of man, and she imposes them all on those who venture out in ships:
Serenity and rage, gravity and vehemence, jocundity and gloom, charity and malice –
The sea is like no man, and yet she is like humanity, boundless as life:
THE ETERNAL SEA.

Walter Kaufmann

Andreas Achenbach: Lighthouse in Ostende

If you would know the age of the earth,

look upon the sea in a storm.

The greyness of the whole immense surface,

the wind-furrows upon the faces of the waves,

the great masses of foam,

tossed about and waving,

like matted white locks,

give to the sea in a gale an appearance of hoary age,

lustreless dull, without gleams,

as though it had been created before light itself.

Joseph Conrad

The mild sou'westerlies of summer barely disturb that vast transatlantic passage between the English Channel and New York, but in winter the ocean is punished by the heaviest of storms. G. Schott

Heavy storms – velocity 11

THE OCEAN AND THE COAST

Beaches and dunes on Sylt Island

FLAT COASTS

On all continents wide stretches of shore line are made up of flat, shallow sandbanks which often give way to sudden, treacherous depths. Wind and waves and the currents of the sea make inroads on the shore, may extend or demolish it. The wide or narrow beaches are broken by dunes and often fringed by forests. But on tropical beaches there are no dunes and sometimes a rich growth of palms borders the waters of the sea.

Receding waves on a flat beach　　　　A smooth beach on the Baltic coast

Rugged cliffs of chalk on the Danish Island of Möen

Cliffs of the Eastern Baltic

BLUFFS

Where the land rises steeply from the sea the surf dashes on a coast of cliffs and soon breaks rocks and boulders from the cliffs that tumble to the beach and cover it to the sea.

A rocky coast

Rocks on Lofoten Island

ROCKY COASTS

Rocky coasts abound and divide the land from the sea and often the sea is broken by shapes like rugged mountains or flat, rocky islands which are, in part, of volcanic origin.

Coast of Islands

The steep red sandstone coast of Helgoland Island

NORTHERN FRIESLAND AROUND 1240

Water and land are rivals – though this may not seem so. When the land is exposed to the sea it is at once menaced by the waters, by the onslaught of the elements. The moving tides, destructive in their force, keep the waters constantly in motion and shift the sea-bed from one place to another. This is a slow process generally, yet at times catastrophies as destructive as earthquakes can be caused by sudden storms and in a single night the appearance of a coastline may change so thoroughly that the new reality no longer tallies with any existing cartography.

Northern Friesland around 1240

It is not known when or how frequently in past millenniums devastating storms flooded the North Sea coast. But old chronicles and maps which date back to a period well after civilization had spread across coastal areas along the southern region of the North Sea bear testament to vast land displacements and geographical changes.

Prior to the year 1300 the present North Friesian Islands were not islands at all, but bound together into a vast area that was crossed by streams and studded with lakes. At that time the coastline of Northern Friesland extended some twenty-five to thirty kilometres to the west and Sylt, Amrum and Fœr and the Halligen Islands were still part of the mainland. Between the High Geest and the North Sea Shore there extended a fertile and densely populated area with villages joined in parishes and on the banks of the river Hever was situated Rungholt, then the greatest trading centre of the northern regions.

At the time of the Vikings, settlers of the North Sea coast had already begun the construction of dams and dykes to guard their homes and their land against the onslaught of the sea. This did not prevent vast geographical changes in the epoch that followed, when, 700 years ago, the North Sea wrought havoc upon the land and ushered in a battle which lasted centuries and caused the destruction of the marshes and all that the settlers had built on them.

In the Middle Ages alone some 85 storms of varying violence flooded this coastal region, though at the time there was no telling whether the destruction was not due as much to the slow but steady sagging of the land as to the storms and the floods.

The greatest of these floods on record in the chronicles occurred in the year 1362 when, between the 15th and 17th of January, giant breakers were driven ashore and drowned the land under torrents of water. With the ascending moon the floods rose higher than anyone recalled and Rungholt perished forever, submerged in those dark waters along with seven villages on the Edomsharde. In a single night more than seven thousand people were drowned in the broiling sea, and when the storms had passed and the floods receded all that remained was a vast seascape with but a few desolate islands. This disaster which has gone down in history as the "grote Manndränke" is said to have caused the death of a hundred thousand people in the region between the Elbe and the Riepen. With each new century the coast of the North Sea was devastated by new storms and new floods, and always there was a vast and tragic loss of human lives, of cattle and of land. The onslaught of the sea never abated, and 300 years later, on the 11th of October 1634, another hurricane caused yet another flood.

If after 1362 a large island had remained, which was later named the "Strand", this island too was now broken up, and of an area which had once encompassed some 50,000 hectares only three smaller islands remained – Nordstrand, Pellworm and Nordstrandischmoor which together made up an area of a mere 9,000 hectares. The dykes were broken in 44 places and they subsequently vanished leaving barely a trace. The stormy sea had torn apart and washed away more than 20,000 hectares of fertile marshland. During the night of this storm some 6,200 among the 8,800 inhabitants of the island were drowned.

Again, in February of 1825, heavy storms over the Halligen Islands swept away many houses and destroyed a fifth of all the wharves.

For thousands of years such disasters were the scourge of the coast-dwellers, nor did the islands fare better in the 20th century – although by then man had learned to build more effective dykes, stronger wharves and houses. They gradually reduced the loss of human lives and land.

With the experience of building better dykes, which was gained over hundreds of years, a new confidence developed that at last an adequate protection against disaster had been found. But then, early in February 1953, the Great Dutch Flood struck large areas along the southern North Sea coast and more than 1,500 people perished in Holland.

1962 – again in February – the estuary of the Elbe rose in a flood that reached as far inland as Hamburg. The tide rose four metres above the normal water level and spread into the very centre of the city. Again, the dykes had proved to be too low and not strong enough, and a city area of 120 square kilometres was flooded, something close to 100,000 people were cut off in Hamburg and 315 fatalities occurred. 225 homes were completely destroyed and another 12,000 damaged. In all, some 20,000 people lost their homes. One had come to feel too safe, too protected by modern technology and had ceased to expect that kind of natural disaster.

SHALLOWS

Waterways in the sandbanks. Narrow channels begin to cross the shallows as gradually the water gathers. Soon they become rivulets of water and soil, then small, then larger streams which form a wide net of waterways.

The sandbanks are caused by the tides which are controlled by cosmic forces. Neither waves nor currents cause the powerful motion of the oceans, but the tides which depend on the gravitational force of the planets, on the pull of the moon and the sun in the main, and on the centrifugal force of the earth-moon constellation. No point on earth remains unaffected by these forces, but only the mobile, fluid sea can make them visible, can reveal the impact these forces have.

Journeying over the oceans, the moon draws up veritable mountains of water. Below the moon, and on the side of the earth that faces the moon, these mountains of water bank up and form huge valleys. Out on the open sea, where no comparisons are possible, the magnitude of the tidal waves that chase across the surface of the earth can barely be assessed. Only when they strike the coast across continental barriers are they revealed as rising and falling tides.

Where the continents face the sea with tall and rocky cliffs the rising and falling tides can be determined only by the level of the water. But where flat and sandy beaches border the mainland, the falling tide often recedes from the coast for vast stretches many kilometres long.

The North Sea with its wide approaches to the Atlantic is naturally much like that ocean, and the tides of the North Sea have determined the character of the landscape. At great speed, spreading south and east, the tides of the Atlantic swing into the North Sea. The high tide which is forced to move around Scotland reaches the southern North Sea twelve hours after the tide that courses through the channel, but finally they both converge and in the course of this may diminish in force or combine into a single tidal wave of greater velocity.

The sea's pulse can be felt in the remotest bay and often far upstream on the rivers. Along the flat beaches between the islands, where the tides rise and fall regularly each six hours, and where the bed of the sea is marked by the currents, a region extends which is neither land nor sea, a landscape that is unique on earth: The shoals.

In the North Sea area wide flat beaches extend from the far north of Jutland to the estuary of the Elbe and from there to the Channel. In this region, from Sylt Island to the estuary of the river Ems, tidal differences of up to four metres have been recorded. These in turn, at the time of low tide, have resulted in a widespread exposure of the sea-bed in northern, western and eastern parts of the Friesian islands – not a long exposure, only a few hours, but long enough to reveal what normally remains hidden. From the green edge of the highlands the ocean's shoals appear as a wide, monotonous expanse, without apparent life, stretching far into the horizon. But he who dares out into this am-

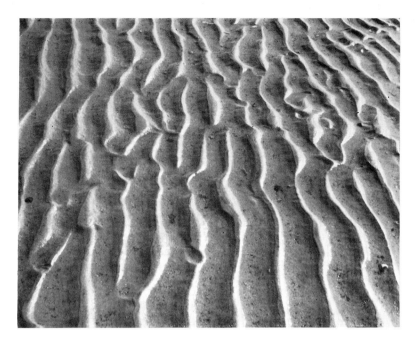 

Various surface formations in the sand of the shoals

phibious world, by foot or by boat, into this treacherous wasteland that lies exposed when the tide is out and is flooded when the tide is in, will discover that the ocean shallows make up a landscape of great richness, and that these swamps are as changeable as the ocean itself.

Although these shallows are essentially a flat expanse with a faint incline toward the sea, the surface is not smooth, nor does it lack contours. A network of waterways, some small, some larger, channels, rivulets, creeks and streams, spreads throughout. In some places, embankments of sand withstand the highest tides, and these embankments are the nesting places of many kinds of birds. Seals abound there and frolic in the sun.

The bed of these shoals is made up of sediments that the sea washes up with each tide and then deposits again. Wind and water give a thousand shapes to the bed of the swamps, a richness of contours like that of a wild and stormy sea. Mostly the waves of the sand are small, like ripples that may seem to glide over the wide expanse of the shoals, ripples that are formed by the currents, by the motion of the sea. Those formed by the currents have long flat surfaces and short steep banks; those formed by the motion of the sea are like waves with valleys and crests – a frozen replica of moving waters. When the motion of the sea and the currents have worked simultaneously or in quick succession, ripples are created that bear evidence of both. Often large areas of the shoals show no variation. The modulation of the sand with all its equal, finely wrought ripples reveals the stark beauty of an elaborate ornament.

At other times, through the interplay of various formative forces, bizzare shapes are formed – forms that resemble fish or skeletons, cocoons, caterpillars, terraces, shapes like leaves, islands of sand that appear to be embraced by the sea and circular holes that are caused by the swirl of the currents and look like the orifices in craters.

The diversity of the forms and their origin are so manifold that the number of variations seems inexhaustible, yet, however diverse these forms are, however diverse their origin, the forces that fashioned them never vary: they are all born of the wind and the sea. The ocean shoals form a wide landscape of timeless change, and always the rhythm of the tides is brought to bear on them. And these swamps, created by the stormy, the windy sea, always manifest the inevitability of change. When the tide flows out and draws the waters back into the depths of the sea, slow at first, then faster and ever faster, the sandbanks and the mudbanks gradually begin to dry. The sea recedes and the shallows look stringy and furrowed. Long after the shallows have been drained of the waters of the sea by the tide, a residue of fluid oozes from the mud and the sandbanks. It is as if each grain of sand flows inside a film of water, and what this film of water can not hold, what it releases, gradually sinks into the crevices. But like the water above the surface of the sand, all this artisian water seeks access to the sea, and where the sand and the mud is more porous it emerges in thin rivulets; and all these rivulets, thousands of them, combine to form creeks. These streams, among other features, invest the swamps with an aura of eeriness, of danger. From the streams the waters of the low tide flow away to the sea, and from the creeks they rise again. They fill up rapidly with the waters of as yet invisible tidal waves, run over and flood the more shallow places of the shoals. The deeper creeks run mostly farther out by the sea, in places of a grey and forlorn solitude which are accessible only to the fishermen, to men on horseback who know the shallows; or else the creeks run in that ominous world, those ominous expanses of a deep and sticky mud into which no one ventures without trepidation, without a feeling of irrepressible horror.

SEA LIFE IN THE
TIDAL REGIONS

At first sight, inexplicable traces reveal themselves in the shoals, these shallows that expand in a leaden silence beneath the sky. Small bubbles penetrate the surface, like the pearls of an oyster they reflect the sunlight for a brief span of time, then burst and vanish. Tiny fountains rise as from tiny wells up to the height of a finger above the sodden soil.
Traces like prehistoric runes are scratched in the

Page 27: Dead shells of sand mussels

Rapacious starfish attacks mussel

Common crab deprived of four legs and its left claw in a battle with rivals or through attacks from sea-gulls

Snails settled on the shell of a large blue mussel

wet surface, and something grey or mud-coloured swirls up as though by a subterranean force – something which, for wide areas, gives the swamps a scarred appearance: innumerable signs of hidden life secreted within the realm of land and sea bear testament to the vast complexity of this amphibian world, and one is taken by the notion that the great secret we call life had its origin only in the water and by the fringe of the sea.

Along the edges of the creeks the fauna of the shoals reveals itself best. Here the animals emerge from the sea-bed to the upper strata of the swamps into areas of greater safety, find protection against dehydration and their natural enemies. Mobile crabs and shrimps, the rapacious starfish and snails all populate the upper strata of the shoals, and only the mussels and worms remain constantly on the sea-bed.

Just below the surface of the swamps there exist untold numbers of small and beautiful heart-shaped

mussels. After days of heavy storms, when the sea and the surf have churned up the shallows, the beaches are often littered with a vast scum of dead and broken shells. Somewhat deeper, in a zone of five to twelve centimetres below the surface, are found the pepper mussels, and of all the mussels which penetrate the surface the so-called sand mussels burrow most deeply down into the sandbanks. Up to 20 centimetres below the surface these sand mussels hide themselves and more than a hundred of them can exist in a single square metre. All mussels depend on the sea water for nutrition and oxygen and so they push long, hose-like pipes through the mud – syphons which explain those tiny fountains that seem to rise mysteriously above the surface. Among the mussels in this tidal region the blue mussel alone thrives on the bed of the sea as well as on rocks and pilings in the surf. Webs of innumerable silken threads bind them together in vast colonies of mussels.

This is the hunting ground of the sea-gulls; here they feast, as from a rich table, hunt mussels and snails, small and larger crabs that have fled the low tides into water holes, puddles and rivulets. Colourful birds called oyster catchers stalk the swamps and dig their long beaks into the mud for worms.

But the beauty of the shoals and all the dangers, too, are fully experienced only if one ventures in a boat far out toward that labyrinth of sandbanks where the seals abound and the sand banks up high and white under the sky like the dunes of an Asian desert. When here in a flood of small rushing waves the tide approaches and the wind blows the foam high over the sandbanks. When the vast ocean surges forward and ever forward in that motion which is due to the cosmic forces, then, as never before, the power of this world may be felt, the impetuous power of this world with all its dark and hidden dangers.

HALLIGEN ISLANDS

Right: Wharves on Nordstrandigmoor

Subsequent double page: The swamps at low tide on the Halligen Islands of Norderoog and Hooge

There are, moreover, in the North such people as the Chauks, some of whom must have been the forefathers of the Friesians. Twice daily and twice nightly the ocean there floods a vast region, puts an end to the feud of the elements, and one wonders whether the region so flooded is part of the sea or the land. The frugal people who live there have erected strange huts which they have placed in such a way that they can withstand the highest floods. The people have thrown up mounds in the sea and built the huts upon them. At the time of the floods the inhabitants resemble sailors, and they seem like shipwrecked survivors when the sea recedes. As the tide draws the fish out to sea they catch them from their huts. These people can never keep cattle, nor have they – as their neighbours have – the use of milk; no, not even game may be enjoyed by them. They can not hunt for game for there is no bush in sight anywhere where game could shelter. From seagrass and long reeds they fashion ropes for the nets with which they fish, and they make some sort of fuel from mud, mud dried by the wind and the sun, which can be burned and used for cooking. They drink nothing but rain water which they catch in wells outside their huts. And these people assert that if they came under the rule of the Romans and thus partook of their culture, they would be reduced to slaves! It is indeed written: Fate spares many in order to punish them.

Plinius the Elder
From: "Naturalis Historia" around 70 A.D.

Around 395 B.C. the Greek explorer Pytheas set out on a memorable journey to which posterity owes the first description of the shoals of the ocean. Pytheas sailed North from the Mediterranean through the Straights of Gibraltar along the Spanish and French coast, and from there around England until he reached the North Sea in the vicinity of the North Friesian coast from where he returned to Greece with stories of the legendary land of Thule.

DECORATIVE WALL PLATE AND DELFT TILES WITH ENGRAVINGS

The life of Holland is that of a country linked with the sea. The land is flat, the skies are high and the sea winds blow across rivers, canals and trenches. Like pawns on a chessboard, medieval towns and townships are scattered across this amphibious territory of water and land. Miles inland from the coast, where these towns and townships are still largely dormant, their cultural and economic life owes much to the pulse of the sea. In the 17th and 18th century Delft tiles with their maritime motifs in a bright cobalt blue became famous across Europe. Frequently, well-known painters fashioned the designs on plates and tiles and in those days it was thought fashionable and a sign of wealth to decorate living rooms with such rarities. Shipowners and sea captains would have the walls of rooms covered entirely with such tiles which, as often as not, were created after their own designs.

RUEGEN ISLAND

Ruegen Island on the south coast of the mid-Baltic is a perfect example of constant coastal changes. The irregular shore of this island, a shore of countless bays, with its beaches of sand and gravel, with its dunes and steep cliffs and shoals, is always shaped anew by the surf and the currents of the sea. Such changes are most apparent where the chalk cliffs tower 130 metres above the sea on the Jasmund peninsula. Here the surf and other forces never cease to make inroads on the soft surface of the chalk, so that in the course of a year layers two metres deep plunge into the sea. And thus fossils are revealed which for millions of years have rested in the depth of the earth – fossils that bear testament to a distant past and which allow us to penetrate a record that might well be termed the diary of the earth.

Maritime life of past centuries – life that frolicked, died, became submerged only to be torn from the white grave and flung onto new beaches by the waves and the storms of new oceans.

Bruno H. Bürgel

FOSSILS FROM EXTINCT OCEANS

Left: Flint stones

Right: Among the fossilized remains of the chalk sea that vanished 75 million years ago the sea urchins are the most remarkable. Flint stone filling of a heart-shaped sea urchin (Cardiaster bicarinatus).

Since time immemorial empty shells of molluscs have been washed ashore. Vanished sandbanks of past epochs, sandbanks since hardened to layers of rock, bear testament to submerged coasts, and what has remained within these rocks – these untold numbers of fossils – reveals an astounding variety of life.
140-million-year-old shell of a cephalopod of the Jura Sea (Proplanulites koenigi).

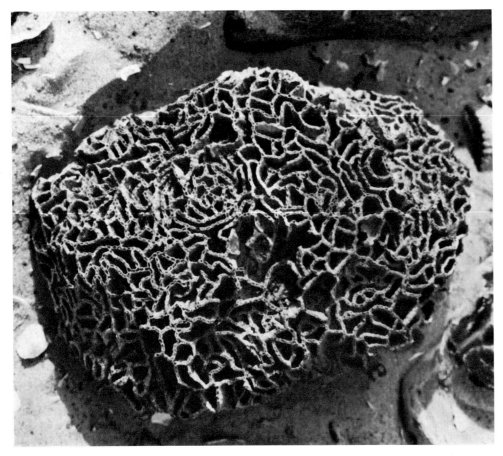

Witness to a tropical sea of a 420 million years ago. A coral (Halysites gotlandica) from the Silur Sea burrowed from the rubble of the Scandinavian glacial period by the force of the surf.

A FISHING VILLAGE IN THE SNOW

And still it snowed. The sea roared, but there was no wind, and it was as though the cliffs and the mountains were listening. When the lights went out and the fishermen's port had gone to sleep, the snow fell heavier still, softly

and without a sound. The snow piled up on the roads, covered roofs and walls with thick layers, and turned stairs into hillocks. Boats and ships stood motionless so that the snow gathered on masts, on stems, on bowsprits, on gunwales. The harbour lights illuminated riggings and lines that were covered with a powder of snow. There they lay, the powerful galeasses and the northland yachts, their riggings adorned with a white lace, and they seemed so slim, so young, there was a whiteness everywhere, something virginal everywhere, and everywhere, in the light of the lamps, something slender emerged that seemed covered as in a bridal gown. These ships that would plough through the sea, vessels that smelled of oil and of tar, now lay here and adorned themselves, and more and more they seemed to glide into a fairy tale. There were the masts that now looked like the spires of churches awaiting the chime of silver bells which would bless them. Now and then, along the bridges, there was a splash in the water – snow had tumbled from the incline of a roof. The white snow and the salty, grey waters of the North Sea merged down there, for a small moment the snow floated, then more and more it turned grey until it sank and vanished.

   Johan Bojer: Lofoten Islands Fishermen

The old lighthouse on the red sand outside the estuary of the Weser. It stands 48 kilometres from Bremerhaven, surrounded by water, on a bank of shifting sand. Since the channel changed considerably in the years after the lighthouse was built in 1885, another more modern lighthouse was erected less than two kilometres further east.

Right: Lightships are always painted red. At a depth of 28 metres lies the Danish lightship of Anholt in the Kattegatt on the southern end of Fyrbanken.

SENTINELS OF THE SEA

Visible from long distances, lighthouses and lightships are dominant markings along the coasts. From sunset to sunrise their burning lights direct ships through the darkness of night to ports and out to the open sea.

The vast coast of the Scandinavian mainland stretches from Cape Lindesnäs to Cape North. Mountains, fjords, gullies and currents of ice are assaulted by the waves of the ocean and torrential rains, are submerged in a dense fog. Some 150,000 islands and cliffs lie in the maze of this seam of the sea. Extending from north to south over 1750 kilometres, the actual coastline of Norway with all bays, fjords and islands measures 20,000 kilometres – the distance from the North pole to the South pole.

Left: Fjord Geiranger at Westland
The currents caused by the ship flow off silently in a perfect symmetry, like finely-cut jewels they reflect the light and the shade, and, glowing with the sheen of silver, they swing toward the bright shores.

MIDNIGHT SUN AT CAPE NORTH

It is midnight. An oppressive silence lies over the sea. The ship glides by without a sound as if moving above the water. The sun shines around the clock and the star of the compass. All sense of time is gone and it seems to the traveller as if it were eternity. And this stanza in Goethe's "Faust" seems in need of a new definition:

>And swift and passing swift revolves
>The gleaming earth in splendour light,
>And Paradisal day dissolves
>In deepest shades of shuddering night.

Polar day and polar night, polar summer, polar winter – all are caused by the rotation of the earth around its own axis, by the movement of the earth around the sun, the incline of the ecliptic,

Detail from a chart of Thule by the Swedish parson Olaf Magnus. The chart dates back to 1535 and shows the trade routes of the tim through the North Sea to Iceland and Norway. A prevalent belief i the menace of sea-monsters caused charts to be marked where such monsters had allegedly been encountered.

the well-nigh spheroid form of our planet. At Cape North the polar summer lasts 75 days in which the sun never sinks below the skyline. But in the polar winter the sun remains always below the horizon; it is the time of the polar night.

Othar, the audacious Viking from the Norwegian province of Halogaland discovered Cape North in the year 870 when he drifted in his dinghy farther and farther out into the North Sea; he circumnavigated the Cape into the White Sea and eventually reached the Kola Peninsula. After that, some 600 years elapsed before the North Cape was again reported on. In search of the famous Northeast passage an English expedition under Stephen Burrough penetrated to Cape North.

For centuries the reef of Cape North was erroneously thought to be the northernmost point of Europe. Since then geographers have agreed that Cape Knivskjellodden, 4 kilometres west of Cape North at the northern latitude of 71°11′8″, actually constitute Europe's northernmost point. Here, on the flat su face of a rock protrusion, a wooden plaque is fixe which bears the inscription: "Europe's End". Sinc both capes are situated on MagERöy Island, though one must seek the northernmost point of the Euro pean mainland at Cape Nordkyn (71°8′1″) betwee the fjords of Lakse and Tanaf.

The 300 metre high summit of Cape North consist of grey-black schist with folds and fissures along th steep fall to the sea. From this steep and craggy fac of rock a horn of rock stands out, a post-glacia needle which points starkly to the sky in the direc tion of Spitzbergen.

Cape North and all of Scandinavia's West and Nort Coast are constantly affected by the Gulfstream tha passes by at a distance of between 50 and 150 kilo metres and which is known as Europe's "hot wate system" that keeps the ocean free of ice.

ICELAND

Land of volcanoes, of glaciers,
lakes and waterfalls,
land of bleakness and of bright
nights, gatherings of the Thing,
and of the Edda.
From the millenium of Iceland's
history and culture resound the
tidings of creation, of the rise and
fall of the gods and of the earth:

In primeval times,
when Ymir reigned:
there was no sand, nor sea
nor waves of salt,
no earth below,
no heaven above,
a bottomless abyss,
and not a blade of grass.

The sun went out,
the land sinks in the sea,
and all the stars
plunge from the sky.
Smoke and fire
spread all about;
A burning heat
ascends to heaven.

I see emerging
once again the
land from the waters,
beginning to green:
waterfalls flow,
and eagles soar,
above the rocks
in search of fish.

BIRTH OF THE NEW ISLAND "SURTSEY" NEAR ICELAND

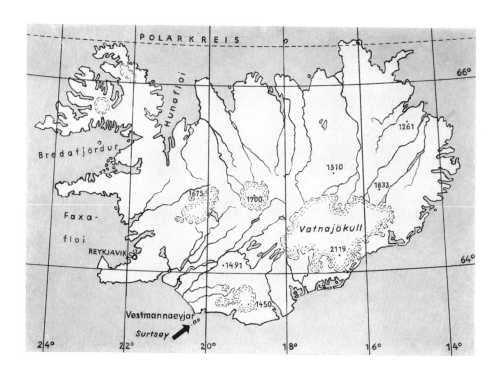

On November 14, 1963 there occurred an event by the South Coast of Iceland which can be likened to a new staging of a play about the creation of the earth: an island was born in the North Atlantic. 3 sea miles WSW before the dark backdrop of the craggy Westmen Islands this event occurred, and for many months it enlivened the Atlantic seascape. One day a change came over the turbulent waters, an appearance that bore no relation to the motion caused by wind and storm. The ocean boiled, the water steamed and cascaded skywards in hot spurts. After that a gigantic black cloud erupted with a great noise, rose 10,000 metres and then was torn by deep red columns of fire. The following night the island was born, and it was named "Surtsey". The volcano under the surface of the water caused the island to grow, and ten days later it had emerged to a height of a hundred metres. In the succeeding months the volcano ejected a fluid of basalt lava which spread over the base of the island and ensured its continued solidity. In the shape of a kind of table mountain the lava built up on top of the island, too, and there a Halemaumau crater boiled for longer than a year. Millions of tons of ashes, lava and rocks shot upwards amid eruptions of fire.
In May 1965, the underwater volcano ceased for the present. From the depths of the ocean it had forced to the surface an island of 2.5 square kilometres which now lies 170 metres above sea-level.

The fjords of the East Coast are cut deeply into the table lands of Greenland. They are as large as the fjords of Norway and as deep. No fjord on earth is larger than the Scoresbysund which penetrates up to 300 kilometres into the mainland and cuts through the rock of the shelf to a depth of 1400 metres. Great glaciers, the largest and most productive existing, flow steadily from the mountains into the fjords, divide and deliver their fragments to the sea. These pieces then travel south in the cold Greenland stream and not till they have moved for several thousand kilometres do the last of such icebergs melt in the warm waters of the Gulfstream.

Sealer in glacial ice in the fjords east of Greenland

THE NORTHERN ICE SEA

Unlike the Arctic Sea, the Northern Ice Sea is contained by the coasts of the mainland, and the only wide approach to the Atlantic is found between Norway and Greenland. The whole of this

Right: Together with other ice-breakers, the "Lenin", largest atomic ice-breaker in the world, opens the northern approaches

north polar basin is packed with ice, an almost unbroken, compact mass of ice, vast areas that only in the short months of summer succumb to streams that make inroads on the southern borders where the ice may melt into larger stretches of water.
Wind and currents cause the motion of masses of pack-ice which, in places, may pile up into high barriers that could squash any vessel.
In 1878/79 the Swedish explorer Nordenskjöld succeeded to navigate the "Vega" through Arctic waters around Europe and Asia along the North European and Siberian coasts. But he had to winter with his ship at the North Coast of Chuck Peninsula. After Nordenskjöld, other notable polar explorers made the voyage, men like Nansen and Amundsen, but they too needed more than a single navigational period, e.g. they had to winter in the ice. As late as 1932 a ship, the "Sibirjakov", finally covered the distance from Arkhangelsk to the Pacific Ocean, some 4000 sea miles in all, without interruption. Since then the Soviet Union spares no effort to open up these northern approaches. Meteorological stations and observation posts were inaugurated so that the motion of ice could be recorded, and powerful ice-breakers were built which today conduct whole convoys or single ships through ice-bound areas. The flagship of this fleet is the ice-breaker "Lenin" which is capable of crushing surfaces of ice more than two metres thick.

Fishing boats in the open sea

CLIFF FORMATION ON THE COAST OF CORNWALL

The extreme south-west coast of England forms a continental bulwark against the swell of the tempestuous North Atlantic. The ever recurring assaults of the waters and their allied elements make inroads upon the land, alter and destroy the structure of the crust of the earth. Due to the erosion caused by the sea, a coast developed that was made up of odd rock formations and often chaotically assembled parts thereof. Ridges, sharp as razors and serrated like saw-blades, and untold other phantastic conglomerations evoke the impression of some kind of dreadful weapons rather than that of a coastline.

Invariably, cliffs and reefs in the sea are a menace to shipping. Woe to the vessel which, caught in a storm and at the mercy of the surge, makes but the briefest contact with such cliffs or reefs. The ship is lost, torn open, split, cut, and wrecked. Reefs tear a steel-plate as easily as a paper bag, they penetrate the ship's belly and bite into stanchions and bulkheads. Many ships grounded on reefs have been captured there for days and weeks, held until the reefs grew weary of their prey, as it were, and let them sink to the bottom of the sea.

BRITTANY

Cornwall in England and Brittany in France each thrust out far into the Atlantic – bulwarks against the sea and guardians of the approaches to Europe's big ports. A craggy rock protrusion forms not only the northern arch of France, but also the beginning or end of the vast sea basin which is the Bay of Biscay. On the Ile de Quessant stands the last or first beacon of Europe. All voyages begin or end here. Seafarers call it Ushant, and in the time of sailing ships no beacon was awaited with more longing, more anxiety and more hope.

It would seem that the coast of Brittany defies all comparisons. On the rest of the continent no greater span between high and low tide has been recorded. Near St. Malo the waters rise and fall fourteen metres. Nor is there in Europe a coast more torn, one that is wilder, more jagged, more deeply cut into the mainland. For thousands of kilometres cliffs abound, and reefs and caves which resound with the roar of the sea, and there are rocky vaults, arches, rocky towers. Relentlessly fierce storms of winter and mighty Atlantic breakers have split and cut the grey schist, red porphyry and yellow sandstone. Over timeless periods the sea has crushed the sandstone

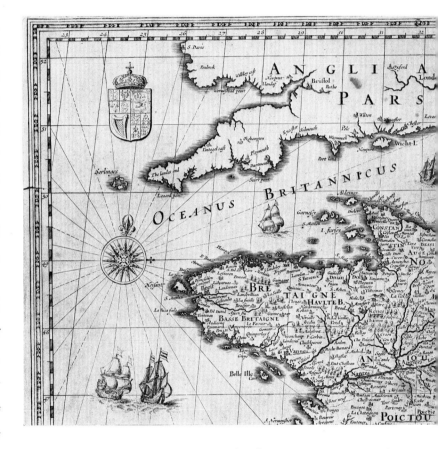

and created lovely beaches of a fine sand among the rock formations in the bays. The sea destroys and the sea builds up. Pointe du Raz in the far west somehow constitutes a geographical symbol for the Brittany peninsula as a whole: flung audaciously far into the sea, a rock protrusion with island cliffs and a lighthouse confronts the Atlantic.

CAPE CARVOEIRO ON THE COAST OF PORTUGAL

Between Cape Finisterre in the North and the high cliffs of St. Vincent in the South of the Iberian peninsula, the West Coast of Europe abounds with such rocky landmarks. The mighty breakers of the Atlantic which surge against them cause cascades of water to rise high above these rocks and, at times, to engulf them. As a pendant to the white lighthouse on top of the cliffs the waves and the surge have created a solitary tower of rock.

COSTA DEL SOL LANDSCAPE

No hectic tourism – and along the Costa del Sol there is much of that – can altogether destroy the enchantment of a deep solitude which one experiences in the course of a day's journey from the Atlantic through the Straits of Gibraltar to the Mediterranean. From the peaks of scorched hills the ruins of ancient fortifications, of walls and towers look out to the sea. There are vineyards and fields in these hills and below there stretches a coast just wide enough for isolated houses and hamlets.

THE GULF OF TUNIS

On this coast lay the ancient metropolis of Carthage which was founded by the Phoenicians 800 years before Christ. For centuries they dominated the western Mediterranean from here and blockaded the Passage of Sicily.

These quiet sandy bays were once the port of ancient Carthage

Small shipyard in La Goulette, today the seaport of Tunis
Right: Gulf of Tunis

For the people of antiquity the Mediterranean was a kind of testing ground for seafarers and maritime explorers. To the ordinary seaman the Mediterranean may not have seemed uniformly propitious. On the whole, though, he did not reject it and found it a sea to be trusted.

The Adriatic coast of Yugoslavia south of Dubrovnik

On the pages following:
Labyrinth of cliffs on the Dalmatian coast
Adriatic breakers at Corcula Island

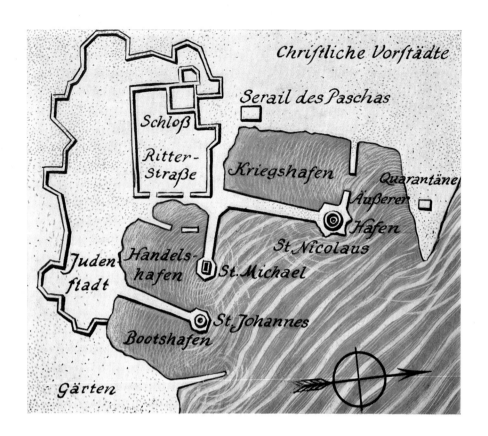

RHODES – MEDITERRANEAN FLOWER ISLAND

Rhodes is the largest of twelve Dodecanese islands in the blue Aegean Sea between Crete and Asia Minor, an island flung farthest to the East within view already of the nearby Turkish coast. The city of Rhodes is situated north-east of a trunk-shaped extension from the shapely island which, after a few miles of treacherous rocks, reaches out to sea.

The town and the island have a rich historical past. In ancient times the Colossus of Rhodes – considered one of the seven wonders of the ancient world – stood on the harbour front in dedication to Helios, the god of the sun. In the pre-Christian era Rhodes was an important trading centre, the Venice of antiquity, as it were. In medieval times the knights of the Johanite Order gave the town its militant appearance. The old part of the town was surrounded by a strong wall, and bastions with tall towers, where, then as now, beacons guided shipping were erected on the pierheads. These days many tourist ships call on Rhodes where they lie at anchor in the bay. The port itself is still frequented by smaller vessels from the eastern Mediterranean – old wooden Kaikis and Peramas with red or brown sails and ornate decorations on the stern. Not far from the port, beyond the azure waters and under an azure sky, the old town-walls appear to burn with crimson glory of roses and hibiscus flowers.

Above: The ancient port of Rhodes
Right: St. Nicholas fortress on the pierhead by the mouth of the harbour and a view of the harbour at Rhodes today

SHELL OF A TRITON

This tropical snail has extended its habitat to the Mediterranean coasts of Europe. The turret of its shell winds down from the apex to the base and the whole shell gleams with that reddish-brown of a twilight sun. It is as if it had been dipped into the reddish glow of a sunset and the colour been burned into the shell. But this harmony of colours and all the gleaming splendour must not hide the fact that the snail is rapacious, an avid hunter of small animals on the sea bottom such as starfish and sea-urchins. The snail emits a poison which dazes its prey before devouring it.

BUDVA ON THE COAST OF YUGOSLAVIA

Many settlements and towns on the Adriatic coast of the Mediterranean are situated on peninsulas that were islands once but which, in the course of time, while whole areas of sand became submerged in water and part of the sea-bottom, turned into protrusions of the mainland. There are records of an Illyrian settlement called Budva which date back to the 4th century B.C. The old town expanded gradually on rocky cliffs. Crooked lanes led through it and a strong wall was built around it as a protection against pirates. Today agaves and acacias grow out of the wall which is slowly disintegrating and crumbling into the Adriatic.

YACHT HARBOUR OF PIRAEUS

The historical waters of the Aegean Sea are today a playground for yachtsmen of the eastern Mediterranean. Where once Greek and Persian ships of war converged in combat and the noise of the battle of Salamis resounded over the blue waters of the Gulf, now not only the sailing boats from Athens are seen to cruise, but sailing vessels from all over the world, deep-sea yachts that brave the sea of Attica with all its cliffs and which, as if in pursuit of a legendary adventure, set their sail in the wake of Ulysses. One of the finest yacht harbours of the Mediterranean, separated from Port Piraeus with its trading ships by a range of low mountains, can be detected in a quiet and secluded bay that extends below the houses of Castello.

THE SUEZ CANAL

Under the red embankment of rock along the African coast by the Gulf of Suez, ships of all the world await their passage through the canal. Today, the Suez Canal leads the three large sea canals in importance. It severs two continents and connects two seas: the Mediterranean and the Indian Ocean. Until the 19th century and since ancient times, a canal connected the Nile and Red Sea, but with the beginning of our era it was no longer viable. The new canal was inaugurated in 1869. From Port Said to Suez, over 161 kilometres, the canal leads mainly through desert wastes until it merges with the Great and Small Bitter Lakes. No bridges cross it and no sluice-gates retard the passage of ships.

ARABIAN FISHING BOATS

Until the Middle Ages the Arabs have been dominant in seafaring across the Mediterranean and the Indian Ocean. Various kinds of dhows were the typical vessels of the Mohammedan tribes. These fast and shapely craft sailed the Red Sea from the Persian Gulf to as far as Zanzibar. Today, such craft are no longer in evidence along the main routes of shipping, but some might still sail out of small Arabian ports on short runs along the coast. From the ports of Africa whole fleets of Arabian fishing boats venture forth upon the sea and, returned with their catch, they can be seen moored along the piers, their heavy brown fishing nets flung across the masts and drying in the sun.

CAPE OF GOOD HOPE

Every day many thousands of ships cross the oceans of the world along passages that are links between the peoples. They transport passengers and cargo. From one port to another, from continent to continent, around the globe. In their bellies of steel these ships carry the heaviest of loads, tons by the thousands, and yet the sea bears them as effortlessly as a feather. The ships ply their course under the moody skies that menace them with all the forces of the elements; they are both sufferers and fighters, and always they must assert themselves.

TERCEIRA OF THE AZORES

Terceira in the North Atlantic is one of nine islands among the Azores. From Graciosa to Santa Maria, over 360 kilometres, these islands are scattered, and they rise from an ocean 4000 metres deep. Pico de Pico, the highest volcano, reaches 2320 metres into the sky.

MIAMI SUNSET IN FLORIDA

The Florida peninsula in southern USA extends 600 kilometres into the sea and so protects the Gulf of Mexico from the Atlantic. The shores of Florida consist of flat sandy beaches with coral reefs and mango-groves. Famous resorts like Miami and Palm Beach are situated on the eastern coast along bays with many lagunes.

Seaview of Havana

THE AMERICAN MEDITERRANEAN

On the fringe of the tropical Atlantic, between North and South America, the ocean becomes a veritable world of wonders. In the West-Indian waters the Gulfstream originates, one of the most marvellous natural phenomena, truly gigantic and unique. Its currents gather in the depths of Florida's Straits and from there the stream flows in its immensity along the American East Coast out into the Atlantic. It warms the shores from Florida to the European coasts.

A vast variety of life frolics in the blue waters, amid the reefs and on the bottom of the sea. Life abounds on all levels of the ocean – from microplankton to the giant shark, from flying fish resembling dragon flies to giant rays that flap about as if with wings. Notable features of the warm seas are tropical corals. For their protection, the tiny coral polyps can solidify matter already dissolved in the water, and that way they form reefs of great length, gigantic castles growing to the surface from a depth of 30 or 50 metres.

Much of the charm of this Atlantic underwater world lies in its variety of forms and the beauty of its mussels and snails. An unbelievable glow of colours suffuses the water, and nowhere in the realm of animals does what remains after the extinction of life long continue to emanate the fascination that the dead shells of molluscs emanate. The manifold and often finely wrought shells enhance the white coral beaches with a colourful and mellow mosaic. The wet sand becomes rich with a gentle glitter of colours, with a glow like jewels in the sunlight. It is as if the brightness of a thousand tiny stars were reflected in the hollows of the shells. And there are shells of a peculiar shape, the shells of the sand dollar, a tropical variety of the sea urchin, cast upon the white sand beaches from the dark underwater world of caves and pools. Considered generally, and more prosaically, an agglomeration of a well-nigh timeless maritime life has been washed upon the beaches here, life that was created and gradually ceased to be.

The chain of Antilles Islands swings in a gentle arc from Cuba to Trinidad: islands that face or are sheltered from the winds. In a sense, these islands and the Yucatan peninsula divide the American Mediterranean into the Gulf of Mexico and the Caribbean Sea.

Tropical vegetation lends the Antilles an appearance of dark green jewels above the blue surface of the ocean. Cuba is regarded the most precious of this group of islands, or the "Pearl of the Antilles". White and sandy is the beach of Veradero. And the surf grinds the coral fragments washed upon the beach more thoroughly than millstones. Water as in an aquarium, bottle-green, suffused with light and clear, washes upon the shores and then flows out again toward the deeper blue of the far distant Gulfstream. Coconut palms sway over the beaches, cast their brown fruit into the sea, and faraway the backdrop of mountains darkens with the fall of blue shadows.

The beach of Ipanema

View of Mount Gavea

GUANABARA BAY AND RIO DE JANEIRO

Few bays on earth have been so generously blessed with beauty as Guanabara Bay. The landscape is grand: mountains soaring skyward, cleavages of rock and deep gorges, dark and silent – yet lovely too, with chains of hills in the sun, forests, gleaming waterfalls and paradisian islands in the waters of the bay. Here we have a world of rough mountains enhanced by the splendour of tropical vegetation and by the blue Atlantic that forces its surging tides from vast distances of thousands of miles in roaring breakers upon the gleaming beaches. The approaches to the bay are fascinating.

A narrow opening, a keyhole as it were, in the rocky coast leads from the Atlantic into the bay where there would be space for all the fleets on earth. Between Pão de Açucar and a rock protrusion on

the Niteroi peninsula the passage is barely a sea mile wide. Then, marvellously, the bay opens out and reveals a splendid panorama. From the mountain range of the Serra de Carioca that reaches close to the coast, some isolated mountains stand out as islands. Rock formations, parts of the mountains and capes protrude into the bays which nestle in this mountainous country. More than a hundred islands with palms and cliffs are scattered in this paradisian bay which already gave shelter to the ships of Magellan on their first voyage around the world.

Rio de Janeiro – Cidade luz – the city of light is a fusion of Portuguese traditions reminiscent of a dreamy Mediterranean port and the hectic life of an American metropolis. Rio must be viewed from the peaks of its mountains to appreciate why this city is considered the most beautiful on earth. It is a cosmopolitan city on the edge of a tropical forest, a city surrounded by mountains, yet always within view of a blue ocean, it is a city fringed with snow-white beaches. The names Copacabana, Ipanema and Leblon spell out six kilometres of Atlantic coastline with a façade of modern hotels, and between the surf and the highway there extends a milky way of beach umbrellas through the hot sand.

There is barely room for the expansion of the city of Rio along the seam of ocean coastline on which it was built. The outer suburbs reach out farther and farther into the valleys of the mountains, and more and more colourful bungalows appear in the dense tropical mountain forest of Tijuca. Apart from those mole hills, those Morros which spring into view suddenly, unexpectedly, the city of Rio reveals two natural attractions which have gained a certain popularity in the world. At any rate, Sugarloaf mountain is reputed to be the best known and most unusual mountain in all of South America, although it rises no more than 305 metres above sea-level. On the far side of the city, across from Sugarloaf, Corcovado – the Humpback – a mountain with a steep and craggy façade of rock, rises from the Jardim Botánico, and from its height it offers an exciting view over Rio and the bay of Guanabara.

The Sugarloaf and the Corcovado at the mouth of Guanabara Bay dominate the panorama of Rio de Janeiro

The coconut palm, a tropical tree, flourishes in the moist-warm monsoons of the Indian Ocean, and probably it originates on the coasts and islands there. No other tree is found on the Polynesian islands or on the uninhabited islands of the Pacific. Its slender trunk with ringshaped traces of fallen leaves is crowned by a bushy, feathery top. Normally the palm grows by the water, swaying in the winds off the sea. Along the sandy beaches the trees form a kind of green wall against the sea. Only seldom are the palms found along the rivers or further inland along mountain ridges. In Indonesian waters one may come across the tops of palm trees drifting in the sea. Storms had broken them from the trunks, had flung them into the surf where the currents had carried them off. Like Portuguese galleys, the tree-tops sail along with some of their branches showing on the surface of the water. Quite often single coconuts cover vast distances, float across the oceans until, finally, overgrown with sea-weed and young mussels, they are flung on some faraway beach — and on such a beach, eventually, new palm trees grow.

The fruit of the coconut palm has become an important ingredient in the nutrition of man. Where they originate they are processed for copra and exported that way. And from that copra oil is extracted, in large refineries, oil that is then turned into an ingredient for the manufacture of cooking fats.

The largest region of islands on earth is found in the tropical waters between the Indian Ocean and the Pacific. It is likely that this chain of islands arose when the world was young and structural changes destroyed all links between Southeast Asia and Australia. A conglomeration of small and larger parts remained as islands, and these parts indicate clearly enough the course of a landbridge that once existed. Many of these islands and some of the more recent Pacific atolls are uninhabited and so small that only a dozen coconut palms or a beacon have room on them.

On the coast of Sumatra

On the following pages:
In the Bay of Along – Gulf of Tonkin
Junk in the waters of Hongkong

In the junk port of Tsingtao · Chinese sailmaker

TROPICAL SEA LIFE

Left: coral   Right: giant mussel
Below: brain coral lodged around giant mussel

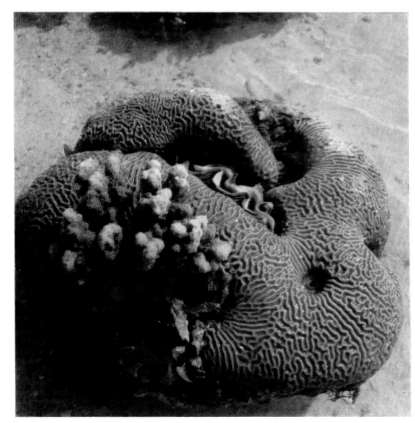

THE AUSTRALIAN COAST

ANTARCTICA

Within the framework of a program for joint exploration the Soviet Union, the USA, Great Britain and other countries have established permanent bases on Antarctica. The sea surrounding this frozen continent is covered with masses of ice and studded with icebergs so that those expeditionary bases are accessible only during the brief months of summer when supply ships make their way there. Headed for the English Antarctica-Base the "John Biscoe" struck trouble between icebound islands and had to be rescued by two American icebreakers.

Pieter Brueghel the Elder: Storm at Sea

THE BIG DEEP

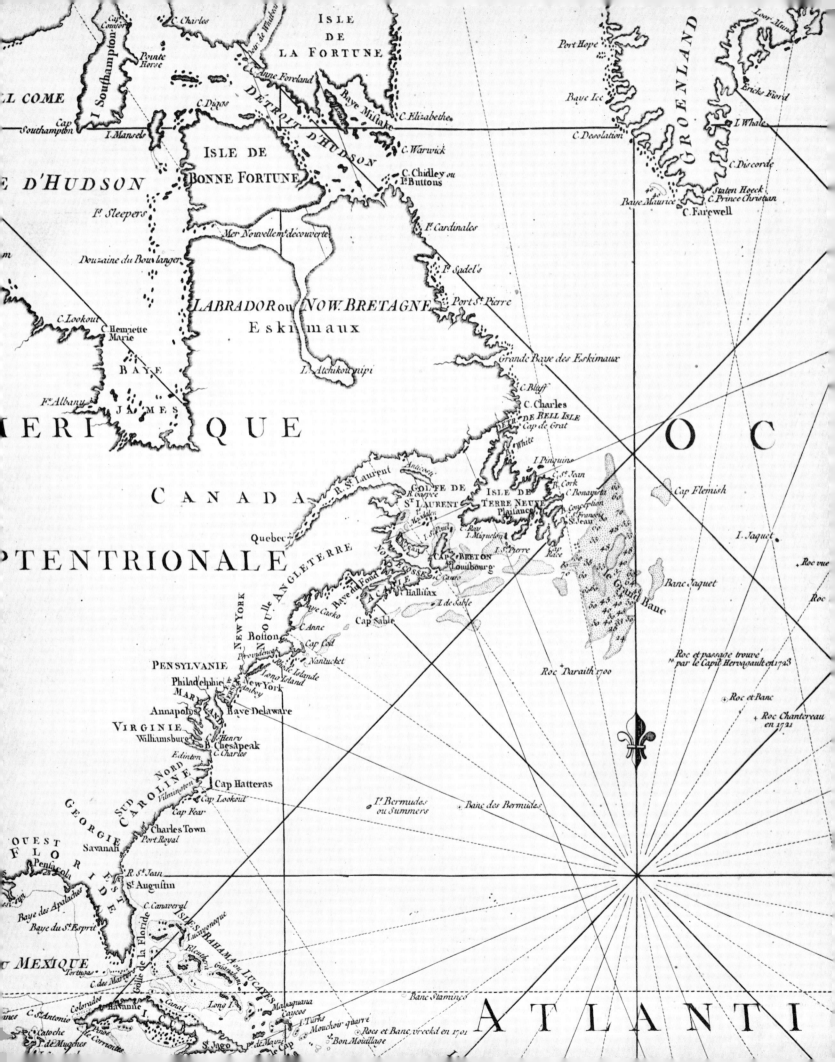

CARGOES

Quinquireme of Nineveh from distant Ophir
Rowing home to haven in sunny Palestine,
With a cargo of ivory,
And apes and peacocks,
Sandalwood, cedarwood, and sweet white wine.

Stately Spanish galleon coming home from Isthmus,
Dipping through the Tropics by the palm-green shores,
With a cargo of diamonds,
Emeralds, amethysts,
Topazes, and cinnamon, and gold moidores.

Dirty British coaster with a salt-caked smoke stack
Butting through the Channel in the mad March days,
With a cargo of Tyne coal,
Road-rail, pig-lead,
Firewood, iron-ware, and cheap tin trays.

John Masefield

As a special complication, the forces of the winds are added to the variety of forces inherent in the sea. Every ship contains a variety of forces. The power of nature, unlike mechanical power, is boundless. A contest is waged continually between these two powers, between a power that becomes effective by virtue of an inexhaustible energy and another that becomes effective by virtue of inventiveness, and this contest is known as shipping. The willpower inherent in the machine is a counterweight against the infinite. But the infinite too is subject to a determined course. There are no blind forces of nature. Man should seek to explore nature and to recognize its laws.

Victor Hugo

STORMY SEA

Next morning the not-yet-subsided sea rolled in long slow billows of mighty bulk, and striving in the Pequod's gurgling track, pushed her on like giants' palms outspread. The strong, unstaggering breeze abounded so that sky and air seemed vast outbellying sails; the whole world boomed before the wind.

Herman Melville

STORMY SEA WITH BREAKERS

Under the influence of an air current over the surface of the water, waves develop. As the winds increase the waves grow. The motion of the waters culminates in a stormy sea. Finally, awesome breakers rise and spread their white crests across the sea.

CROSS SEA

If suddenly, or within a short time span, a wind that has already put the sea in motion changes its direction, new waves are formed that follow that direction, while the old waves remain on course. Thus varying systems of waves run counter to each other and a wild cross sea is created, a menace to ships, since it is difficult to avoid breakers that run athwart.

GROUND SWELL

After the wind subsides the motion of the waves continues across the surface of the sea. A swell develops. The waves lose their pointed crests and become rounder and flatter. Simultaneously, the length of the waves increases. In a becalmed sea the swell may continue for days on end and may spread over vast areas of sea.

WAVES

In a storm the seascape is in motion from horizon to horizon, deep waves carry the cold whiteness of turbulent, wind-tossed breakers. The mountainous grey-green, silver or black waves rise and then roll away under the moody sky. The waves are huge and the valleys between them dark, steep, and deep as caves. The up and down motion of the waves is incessant, and the sea appears to breathe under the waves.

All phenomena of nature obey laws which determine the course of developments. In a calm, the surface of the water is shiny as a mirror and as smooth. The slightest current of air causes small wavelets, cat-paws as the seaman calls them, to hurry as dark shadows over the water. These constant air currents emit a pressure to which the surface of the water reacts like the skin of a drum.

Thus, the first small waves are created. With the prevailing wind the waves continue to grow, to fall more heavily which, in itself, influences the course of their motion. There is a relation between the velocity of the wind and the height, length and motion of the waves. The force of the wind is transmitted to the water directly and through the surface friction created by the air currents.

The eye is deceived when it beholds the waves of the sea which appear to be moving masses of water. In reality, only the outer forms of the waves proceed steadily forward, and what we see is no forward thrust, no powerful motion of the water itself. The motion of the waves constitutes no simultaneous motion of the water. And this is proven by the bubbly foam on the crest of the waves, for this foam remains almost stationary and does not follow the waves, even when these attain the velocity of the wind. The same applies to objects cast overboard and left to float in the sea. Birth and growth of a wave is an intricate process which has occupied the minds of learned men for centuries. Already Leonardo da Vinci concerned himself with the way the waves of the sea originate, and even today such inquiries have not been exhausted and much remains unsolved.

Within a chain of waves the water particles themselves barely proceed, remain almost stationary. At the same time though, while the waves swing up and down, they follow a circular or elliptical course. This course determines the size, height and length of an entire wave. Underwater photography has revealed that in the valley of a wave the particles of water move against the wind and against the direction of the wave; within the ascending wave though, they move in its direction. This orbital movement, as the circular motion of water particles within a wave is known, causes the wave to advance, makes such advancement possible. While the water particles describe a circle within each wave, the wave itself has advanced by its own length. In the crest

of a wave the water particles come under the direct influence of the winds, are consistently forced in the direction the wave is taking and so never return to their original point of departure. Thus, what we see rolling along as a giant wave is not the water itself, but only the motion of a wave. If this were otherwise, if the water particles themselves moved forward, no shipping would be possible.

The wind which causes the wave also determines its course. During the inception of a storm the wave gains quickly in height, becomes short and steep. To an extent all waves absorb the wind's energy. They grow in relation to the force of the storm, and as they grow their velocity increases. Their form becomes more pointed and steeper until they lose stability. Their crests keel over: and thus the ominous breakers arise.

The size of a wave in the open sea is dependent on the force of the wind, on its duration and constancy, both in the direction the wind blows and the intensity with which it works upon the surface of the sea. This intensity is known as "fetch" or "stroke length". Thus the wave grows with the fetch of the wind. In the Baltic, for instance, no extraordinary waves have ever been recorded, since a stroke length of at least five hundred sea miles is required for the development of heights above the normal. Moreover, the height of a wave is also dependent on the depth of the water which must be at least as deep as the wave is long if it is to unfold to the full – in other words, flat waters retard the development of waves. Even in the fiercest storms the height of the waves in the Baltic seldom exceed five metres. This does not mean that the swell in the Baltic is less dangerous than elsewhere. The very shallowness of the Baltic and the Kattegat can cause the sea to become extremely choppy.

In the larger area of the North Sea waves can attain a height of six metres in the South, a height of nine metres in the North. To develop to the fullest, big ocean waves require a stroke length of up to a thousand miles. Depending on the force of the wind they may reach a height of eight to twelve metres. Scientists and seamen seldom agree on the absolute height of ocean waves. Observations on board the American vessel "Ramapo" have become famous to a degree. In 1933, en route from Manila to San Diego, the "Ramapo" was caught in a Pacific storm, a hurricane lasting seven days. The ocean rose to the height of mountains. A giant wave was recorded which, after calculations, was estimated to be 32 metres high. No greater height had ever been logged. As the storm abates and finally ceases, the waves lose momentum and height. Their latent energy though, gives them greater length and more speed, for length and speed are interdependent. The short, steep waves created in the storm give way to a flatter, longer and more even swell which spreads

Atlantic storm waves with breakers in winter

Following double page: Storm
The surface of the ocean is covered with foam, and mighty breakers crown the waves

all over the ocean. Since the Second World War wave-meters have been placed on the seabottom along many coasts and these register the waves that run above them. In that way, waves have been recorded at Land's End on the coast of Cornwall which originated in the South Atlantic, moved through the Atlantic and reached the British coast from as far away as Cape Horn – a journey of over 6000 sea miles.

In the lesser seas the swell always indicates an abating storm, whereas the swell along an ocean coast is, as often as not, a harbinger of foul weather.

As a wave surges toward the coast from across the ocean and reaches areas of lower depths, it soon encounters retardation due to the rising sea-bed. As soon as the depth of the water under the wave measures less than half the length of the wave, the form of the wave alters noticeably. The friction from the rising sea-bed breaks the velocity of the water particles deeper down, while the higher water particles continue on course at much the same speed. The crest precedes the wave. Finally, a steep and hollow wave confronts the shore until it crashes and dissolves with the sound of thunder. Only now a true displacement of water occurs within the wave. Water and foam pour out over the sand and dissolve in it.

At wind velocity 6, according to the Beaufort scale, the swell in the open ocean has reached a stage where any ship on course against this swell begins "to labour and ship water", as the seaman puts it. The bow is lifted by the wave; and as the swell surges off beneath the ship, it dips and plunges into the next valley of water. The length of the wave, the size and speed of the ship determine the force of the impact. The forecastle-head absorbs the force of the oncoming wave, a momentary tremour is felt from stem to stern, and it seems as if the progress of the ship has been retarded for an instant, but presently the ship overcomes the impact of the wave and continues on its way. When the waves pound against the ship's sides more forcefully, an explosive boom fills the air, and vibrations pass all the way through the hull.

When that happens, seconds after such a boom, all of the bow is awash with a cloud of white, foaming water — water, the force of which can damage the

superstructure and shift the heaviest deck cargo. Clouds of spray beat up against the bridge and obscure the view of the man on the wheel. In a heavy sea the ship may remain submerged for long seconds, ominous seconds ... until at last the bow emerges again. Torrents of green water cascade overboard, the ship seems like a giant in tears, and for a long time afterwards the decks are awash with water seeking an exit. Bad weather is always a test of the seaworthiness of ships. The force of the onrushing waves and the velocity of the ship invariably call for special nautical manœuvres, but there may be occasions when circumstances are so adverse that all such manœuvres fail and the ship succumbs to the powers which resist it.

TRADE WINDS AND MONSOONS

There is no part of the world of coasts, continents, oceans, seas, straits, capes, and islands, which is not under the sway of a reigning wind, the sovereign of its typical weather. The wind rules the aspects of the sky and the action of the sea. But no wind rules unchallenged his realm of land and water. As with the kingdoms of the earth, there are regions more turbulent than others.

Joseph Conrad

THE TRADE WINDS

Left: The day before sunrise
At dawn the black sky of the night and the water become suffused with the colour of blue. As yet the sun is below the horizon, but the clouds nearby are aglow with fire. Only the white clouds of the trade winds, appearing softly out of the night sky, sail by like transient shadows.

Steadily the trade winds blow across the tropical and subtropical regions of the oceans. They are the actual and constant sovereigns of the sea. For centuries no other sea winds have affected the courage and the hope of so many men. In the Middle Ages the trade winds speeded the ships of audacious sailors through voyages of exploration, just as later they speeded those four-mast-barques which were heavy with cargo. In the time of sailing ships the trade winds were of such dominant importance that the development of shipping, economy and culture depended on them. Yet the trade wind is not such a violent despot as the westerlies that storm across the North Atlantic in winter or the "Screaming Forties" that punish the southern oceans of the world. The trade wind is in accord with those who rely on it.

The origin and presence of trade winds are related to the dynamic processes in the circuit of the great atmospheric air currents. They emerge and exist like all other winds which arise due to varying pressures of air over the earth, except that their circuit is not confined to small and narrow geographic areas, but spreads over wide oceanic expanses between the continents. The rotation of the earth diverts the trade winds from their original course toward the equator. In the northern hemisphere, therefore, they blow from the Northeast and in the southern hemisphere from the Southeast. The trade winds occur in all three oceans. They are less moody than all other winds on earth and throughout the year they blow steadily from one direction with a barely changing force. The sky crossed by the trade wind remains as fair as the wind itself. Occasionally, only the scattering of small white cumulus clouds in the blue vault of the sky gathers into a loose, transparent blanket. The air currents of the northeastern and southeastern trade winds passing toward the higher latitudes of the equator never merge, but are constantly separated by a zone of calm at the equator where gradually they lose their force and fade.

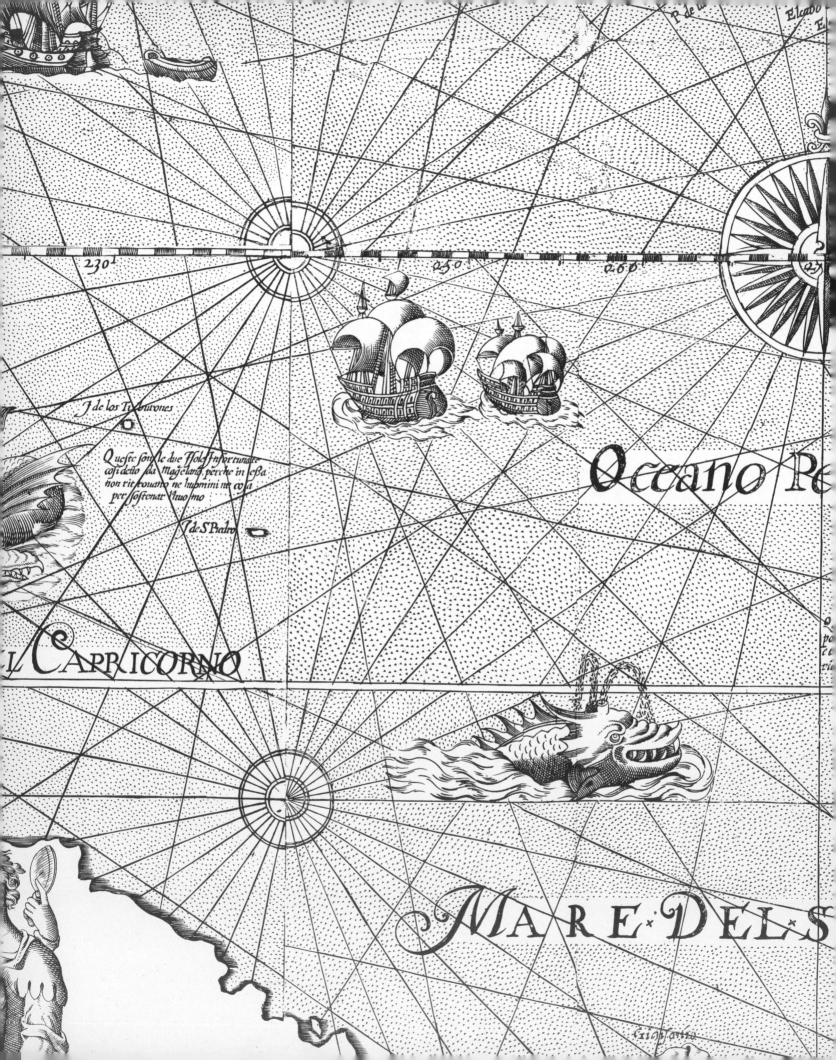

THE MONSOONS

On the following pages:
Monsoon atmosphere in the Makassar Straits between Kalimantan and Sulawesi. A mighty thundercloud rises on the horizon and spreads across the sky with uncanny speed.

On the whole, the monsoons of the Indian Ocean are not as steady and reliable travel companions as the trade winds of the Atlantic. They are winds that change with the weather and alter their direction almost every six months. The monsoon is a grander reflection of the small and playful country and sea winds which, in the summer months, stroke the coasts of the minor seas.

In their repetitive circuit these winds blow only in the daytime from the sea to the land. They abate at sunset and reverse direction through the night. The system of monsoons obeys a similar rhythm in the course of a year. Always a monsoon retains much the same direction for six months, then, as climatic conditions alter, so the direction of the monsoon alters in the course of the period which follows.

The monsoons dominate the northern Indian Ocean. From the equator to the vast bays of the Arabic and Bengal Seas they shape the contours of this ocean. In the winter months, when elsewhere storms abound, the monsoon may be likened to a good friend, peacefully and pleasantly blowing with moderate force from the Northeast, while fair and dry weather prevails and the ocean lies under a sky of cobalt-blue, much as though a trade wind were blowing. But in the months of spring, when gradually the mighty continent of Asia takes on warmth, the monsoon, out at sea, prepares for a reversal. Wind and clouds enter a rebellion. The sea takes on a grey appearance and white-crested waves chase over the surface. The sky, pleasant throughout the winter months, acquires an ominous look. Black rainclouds emerge and expand three, four, even five thousand metres into the sky. The outpour that follows can not be called rain, it is a deluge of water-torrents that streams, pours down – perhaps, in bygone times, such downpours filled the dried-out beds of oceans.

The relation between man and the monsoon indicates that it is a wind that has been known for a long time. Even in the days of antiquity the Arabs had arranged their voyages according to the monsoons. They did not brave the Indian Ocean with their ships, but remained in port until the monsoons of summer had abated. Even today, navigational guides recommend that in steamship voyages across the Indian Ocean the force of the monsoons should not be underrated.

MAN AND THE SEA

Sailors on the Marsrah clearing spider bands

Double page following:
Notions of past centuries
The Flying Dutchman

Seafaring in ancient Egypt

The forefathers of Ulysses, paragon of all seafaring men, appear to have sailed the waters of the Near East before the Trojan War in the 11th century B.C. and it is likely that they also explored the western Mediterranean. However, what we know of the Greek hero's forefathers is scanty. Only fragmentary records remain.

In the past, and more recently, repeated attempts have been made to reconstruct the probable travel routes of the sea heroes of antiquity from Homer's poetic work "The Odyssey", generally considered a literary account of the travels of Ulysses. But since Homer's description could apply to many places and a variety of geographical locations, Ulysses' course can not be ascertained unequivocally. Even when the navigational possibilities of the time are taken into account, no absolute reconstruction of Ulysses' travels is possible. But it is probably that Ulysses was the first sailor who navigated by the stars, by the North Star in particular, and it can be assumed that his navigational skill left room for no anxiety in his mind while he remained at sea for days, even weeks; and thus it can be equally assumed that he actually reached the Columns of Hercules and the Rock of Calpe.

Ulysses' probable course on his voyage through the Mediterranean after the Trojan War (around 1200 B.C.), a reconstruction based on the research of the Englishman Ernle Bradford.

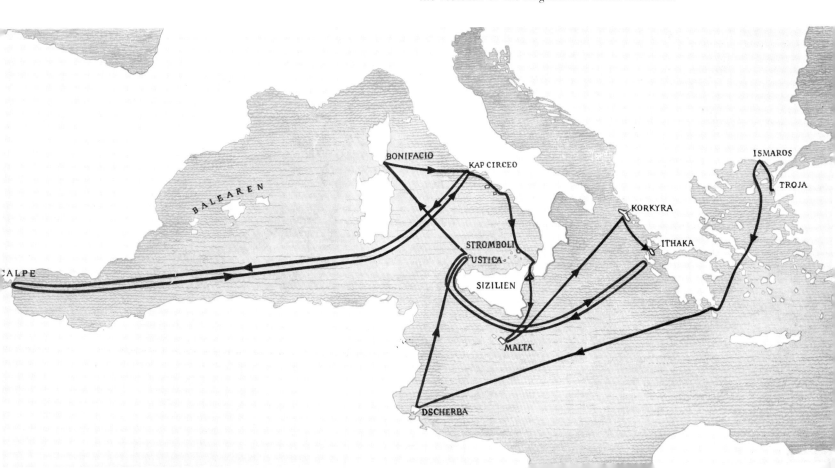

RELIEF PLAQUE OF THE ASSYRIAN FLEET AROUND 600 B.C.

In antiquity, the nautical trade in Asia Minor went beyond Port Euphrat and Port Tigris as far as the Persian Gulf, the Arabian Sea and the Indus. King Sanherib of Assyria is supposed to have manned his fleet with Phoenician crews, and the ships were built by Phoenicians. Apart from the steering gear, the bigger ships carried masts and sails. Whether the ram-ships were exclusively ships of war, can not be ascertained; trading vessels, too, were fortified with bulwarks and shields.

THE VIKINGS

From the history of the seafarers the Vikings emerge relatively late. Hailing from the Atlantic and North Sea Coast of Scandinavia, they were reputed to be audacious sailors. What the Vikings accomplished at sea between the years 800 and 1100 relegates nautical feats of antiquity to second place. Their open vessels had sails and could be rowed by hand. Stem- and stern-posts were finely arched upwards above the waterline and were richly decorated. With these seaworthy ships the Vikings were the first to sail and to conquer the treacherous and stormy North Atlantic. Their enterprise took them to Iceland, they discovered Greenland and settled there, and on the far side of the Atlantic they reached the North-American coast of Labrador. But their settlements there and in Greenland perished after a few centuries and were forgotten. The Vikings were sailors and not ambitious colonizers. For their ocean journeys they developed astronomical navigation. The North Star, the magnet and the sundial were basic and elementary to their primitive methods of navigation.

THE FEAT OF COLUMBUS

The discovery of America by Columbus is considered to this day the most popular event in the history of exploration. Without doubt, though, Columbus was not the first man to step on American soil. The Vikings had reached that continent five hundred years before him. When the adventurous Italian rediscovered America in 1492 the time was ripe for such a feat. The wind soon obliterated the imprint of his boots on the sandy beach of Guanahani Island in the Bahamas, but that morning of the 12th of October the steps of Columbus inaugurated an epoch of dominant importance for the development of the whole of Europe.

Columbus had meant to realize an old dream – he wanted to find a sea route to India by sailing westward across the Atlantic. He lost his bearings. The islands he discovered were not part of India, the land of his dreams. He had discovered a new world of which he could have had no inkling, for

he believed until he died that he had reached India located westward beyond the Atlantic.

Columbus' life and accomplishment are controversial. Nothing in his past is unequivocally proven. He was torn by an obsession for adventure and power. Gold was the magic word of his time, and he coveted gold. He promised to conquer untold riches for the Queen of Spain. He was the son of a poor weaver from Genoa. After desperate solicitations and many setbacks his luck finally turned when the Spanish Court provided him with three ungainly ships for his enterprise, appointed him to the admirality and made him Governor and Viceroy of whatever lands he might discover. With such a contract he set sail.

A map by the Florentine astronomer Toscanelli was instrumental in the plans of Columbus. But Toscanelli's speculations were wrong. Although more than a thousand years had passed, Toscanelli still relied on the calculations and estimates of the Alexandrian scientist Ptolemaeus. The map which resulted gave readings of the distances on earth which were much too small. But this very error of Toscanelli gave wings to the intentions of Columbus and eventually led to a feat in the history of exploration which is second to none in importance.

The Atlantic journey of Columbus with three outdated ships in poor shape did not lack daring in that it led into the unknown. The whole enterprise could be based on nothing more than speculations and legends. Only Madeira, the Canary Islands and the Azores were known and settled at the time, but what lay beyond nobody knew. The crossing of the Atlantic in itself was no exceptional nautical feat and stood in no comparison to the voyages of the Portuguese or Magellan's audacious sailing around the globe.

On the 3rd of August 1492 the "Santa Maria", the "Pinta" and the "Niña" embarked from Port Palos, but three days later the "Pinta" damaged her rudder and could not be manoeuvred. Columbus was forced to make for Port La Gomera on the Canaries where the "Pinta" was put into dock for re-

pairs on the rudder. On the 6th of September he set out to sea again, and then the actual voyage into the unknown began. They traversed the regions of the trade winds with the weather constantly favourable, except for occasional calms. The ships of Columbus sailed with the currents and no storms retarded them. On the 25th of September he wrote in his diary: "The sea remained so calm and so smooth that many sailors jumped overboard for a swim." Only once, a day before landfall, he mentioned choppy seas. Later then, after a further 36 days at sea, he reached the islands in the West, and on the morning of the 12th of October Guanahani was sighted. The whole of the exploit had demanded no sacrifice and was terminated without a mishap of any consequence.

Nonetheless it follows that the historical recognition of his accomplishment must arise from the unwavering tenacity with which Columbus pursued his plans for fourteen years. Throughout that time he fought for his idea. No one believed him, even most seamen had only scepticism for his plans. He returned home from his journey of discovery as an admiral and was subsequently knighted. Columbus undertook four more journeys into the regions of Central America. But his fate proved inconstant. Fame and glory and the chains of prison were his lot. He died a lonely and deserted man in the year of 1506 at Valladolid.

JAMES COOK

James Cook was the last great seafarer in the history of world discoveries. He was born in England in 1728. He was basically a just man and his approach to the natives of newly discovered regions was marked by benevolence. For all that, his fate corresponds to that of many fearless, courageous seafarers before him: In the course of his last journey around the world in the year 1779, Cook was murdered under tragic circumstances by natives of Hawaii.

Thrice Cook sailed around the world on various ships and no captain before him undertook more distant journeys. Admittedly, Cook accomplished this 250 years after the first circumnavigation of the globe by Magellan and his mate Del Cano. Although the surface of the earth was largely known already, certain areas in the Southern Hemisphere had not yet been geographically recorded. Scientifically though, everything still hovered in uncertainty. Besides, many discoveries that had been made in the centuries before Cook had again receded into obscurity, or else, geographical calculations dating back to the 16th and 17th century had been so incorrect and misleading that many islands and parts of countries were never found again. Often too, newly discovered areas were kept secret.

Furthermore, and this was the most burning question in Cook's time, since antiquity the supposition had not been disproved that some vast continent existed deep down in the southern latitudes – the earth, so it was reasoned, was bound to have a counterweight in the South that would balance out the continents of the Northern Hemisphere.

All of Cook's voyages were more than enterprises of discovery or conquest on behalf of the Crown of England, he aimed at more than Vasco da Gama, Columbus, Magellan and many others whose chief concern had been the access to new trade routes and colonies. Cook was a progressive man whose mind was open to the spirit of the 18th century, and in the main he planned and undertook his journeys with the aim of a rational investigation of the problems of seafaring and geography.

He mapped out newly discovered coasts on the basis of exact surveys and researched compass deviations caused by the magnetism of ships. He insisted on having scientists on his expeditions and was ever ready to make use of their botanical and medical discoveries. Thus, he is known to have supplied his crews with large quantities of lemons and onions as a preventive measure against scurvy.

Cook lived at the turn of two centuries, at a time when a new rationalist attitude based on exact sciences began to supersede an approach based on legends and hearsay dating back to the Greeks.

A new era of development, with Cook as a successful initiator, had begun.

In 1768–71, on his first voyage around the world, Cook discovered the Society Islands in the Pacific. He sailed around New Zealand and found the passage which, later named after him, gives New Zealand the aspect of a double island. Further westward he discovered Australia's hitherto unknown East Coast which he explored in a northern direction as far as the treacherous waters of the Great Barrier Reef. Here he rediscovered the Torres Straits kept secret by the Spaniards since their discovery by Torres in 1606.

Cook's second voyage from 1772–75 turned out to be his most successful. He was the first to circumnavigate the unknown ocean which stretches around the world at high latitudes on the Southern Hemisphere. In search of the legendary southern continent he reached the southernmost width at 70°10′, but was prevented from penetrating farther South by icebergs and barriers of ice. The voyage was a tremendous achievement in that it gave science the proof of the non-existence of this speculative southern continent where people were supposed to live. Cook surmised though, and realized correctly, that beyond the icefields all around the South Pole a block of land must exist which was concealed by ice. On his third voyage around the world Cook set out for the Pacific again and discovered the Hawaii Islands. Then he steered northward along America's hitherto unknown West Coast as far as Alaska and passed through the Bering Straits. He penetrated the northern Ice Sea as far as 70°41′ northern latitude and finally abandoned his search for a northwestern passage when he encountered a boundary of pack-ice. He reversed course and headed for the Pacific again, only to meet his death in the Hawaii Islands.

THE SEA – LIFE INCARNATE

Why is almost every robust healthy boy with a robust healthy soul in him, at some time or other crazy to go to sea?
Why upon your first voyage as a passenger, did you yourself feel such a mystical vibration, when first told that you and your ship were now out of sight of land? Why did the old Persians hold the sea holy? Why did the Greeks give it a separate deity, an own brother to Jove? Surely all this is not without meaning. And still deeper the meaning of that story of Narcissus who, because he could not grasp the tormenting, mild image he saw in the fountain, plunged into it and was drowned. But the same image, we ourselves see in all rivers and oceans. It is the image of the ungraspable phantom of life; and this is the key to it all.

Herman Melville

FROM: THE RIME OF THE ANCIENT MARINER

The fair breeze blew, the white foam flew,
The furrow followed free;
We were the first that ever burst
Into that silent sea.

Down dropt the breeze, the sail dropt down,
'Twas sad as sad could be;
And we did speak only to break
The silence of the sea!

All in a hot and copper sky,
The bloody Sun, at noon,
Right up above the mast did stand,
No bigger than the Moon.

Day after day, day after day,
We stuck, nor breath nor motion;
As idle as a painted ship
Upon a painted ocean.

Water, water everywhere,
And all the boards did shrink;
Water, water, everywhere
Nor any drop to drink.

Samuel Taylor Coleridge

Four-mast-barque in the swell of the Atlantic near the equator

Sailors on a motorship caring for the heavy derrick shrouds

Dock workers belaying hawser of deep sea ship to mooring bollard

The men who moor ships are specialists among waterfront workers. The mooring of overseas vessels is an art requiring skill and care. When a ship is about to berth pier-men are ready to seize the heaving lines with which they take up the heavy hawsers to tie them around a post.

BIRTHPLACE OF SHIPS

Warnow Docks, Warnemünde. On four slipways under a cable crane installation 60 metres high, ships are assembled in sections. A fast freighter, a motorship of 12,000 tons and two other vessels are here under construction.

FAST FREIGHTER

Class Hammonia, HAPAG, Hamburg
Docks: Blohm & Voss AG, Hamburg
Overall length   164.31 metres
Width   22.50 metres
Capacity   12,544 tons
Engine Power   18,900 HP
Cruising Speed   21   knots

Ships of this type are among the world's most modern. Only economic considerations dictated their designing. Their building costs and running expenses are high, but they are faster than normal ships and capable of more round trips in a given time. Only large shipping companies with well-established cargo runs can afford to put such expensive ships on order and into service. The Deutsche Seereederei Rostock ordered a series of such freighters from the Warnow Docks in Warnemünde.

PASSENGER SHIP

The "Raffaelo" of the Italia Line in Genoa – 45,933 GRT and a cruising speed of 26.5 knots – belongs to the finest ships of the last years. She runs between the Mediterranean and the USA. 12 decks in all afford 1775 passengers every possible comfort.

Page 178: Ships taken out of service.
A small coastal schooner berthes in Wismar.
Page 179: Sea and river craft in Maashaven, Rotterdam
Page 180/181: In New York harbour

SUPER TANKER

"Tokyo Maru": Tokyo Tanker Company
Overall length   306.5   metres
Width   47.5   metres
Loading depth   16   metres (Water displacement)
Capacity   152,400 tons
Cruising Speed   16   knots
Engine Power   28,000   HP

The loading capacity of tankers has been increased enormously. The "Tokyo Maru" soon ceased to be the largest ship in the world; the "Mitsu Maru" succeeded her with 200,000 tons. A tanker with a capacity of 300,000 tons was put into service by the Japanese in 1968.

Captain

Boatswain

Third Mate

182

Sailor

Cook

Ship's Engineer

THE DEVELOPMENT OF SEA CHARTS

When in prehistoric times the first courageous seafarers braved the open ocean and their home shore receded behind the horizon and vanished from their view, they must have been oppressed by the thought of how to find their way back. This soon led to the necessity to enter the shape of coasts and the position of newly discovered islands on some kind of chart.

Among the oldest of these charts are those, made of small sticks and the ribs of palm leaves, which the natives of Pacific islands designed. Shells and small pieces of coral marked islands and atolls. Seasoned seafarers of the time were able to read routes and distances and the direction of winds and currents from these charts.

What we know of the developments of sea charts among western peoples dates back to the Middle Ages. But it may be assumed that the Greeks began to chart the coastal regions of the eastern Mediterranean around the beginning of our era. Although astronomy and mathematics had already reached a high level in antiquity, the first actual sea charts, so-called portulans, lacked a mathematical or cartographical basis. Although these charts indicated longitudes and latitudes and the shape of coastlines, none of this was based on special surveys, but on hearsay, diaries and the sketches of seafarers. Outstanding coastal locations were designated by symbols, however it is noteworthy that the portulans give no reading of depths, although such readings are essential for navigation and the plummet was known and in use before our era.

With the discovery of America in 1492 and the first circumnavigation of the world 30 years later came a growing interest in usable sea charts. Cartographers began to work scientifically. The geographers fixed a zero-meridian and the cartographers entered all continents, the known and the unknown, into a grid projection. The Pope called for a demarcation line west of the Azores through the Atlantic and thus divided the world into two halves. Fantastic depictions of sea monsters devouring ships decorated these charts. It soon became apparent though, that all these charts contained a variety of errors.

The charts revealed distortions of the coastal regions entered on them and their graduation was too inaccurate to navigate by them. Any chart is useful only to the extent that a course can be plotted and a compass direction between two given points determined thereby. Furthermore, such a chart must allow for the entry of an intended course in a straight line that will cut all longitudinals at an angle that corresponds with the actual angle on earth. The flat chart approached this requirement, although its design still allowed for inaccuracies.

In 1569 Gerhard Mercator published his epoch-making "World Chart for the Use of Seamen". As

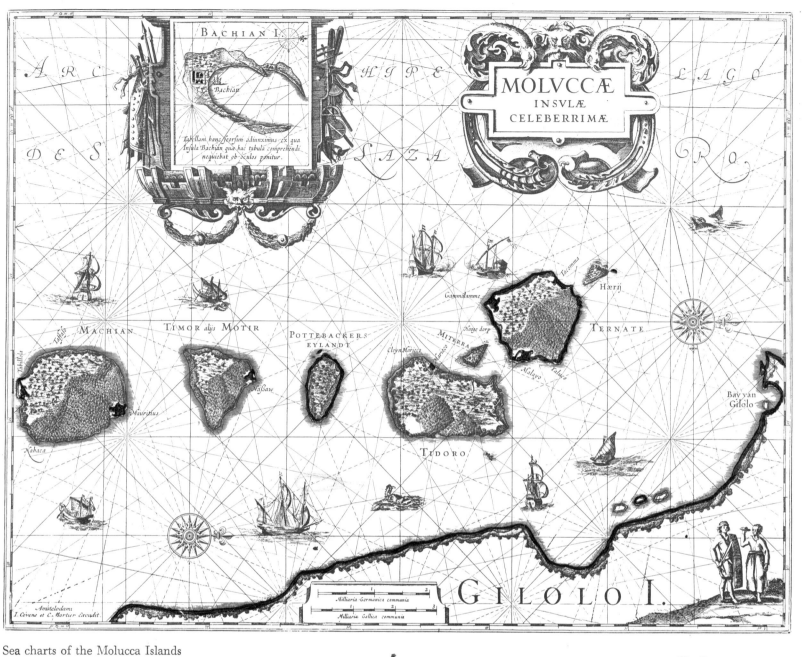

Sea charts of the Molucca Islands

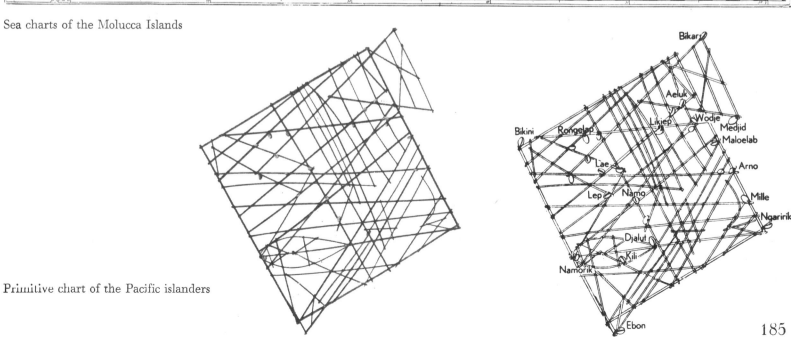

Primitive chart of the Pacific islanders

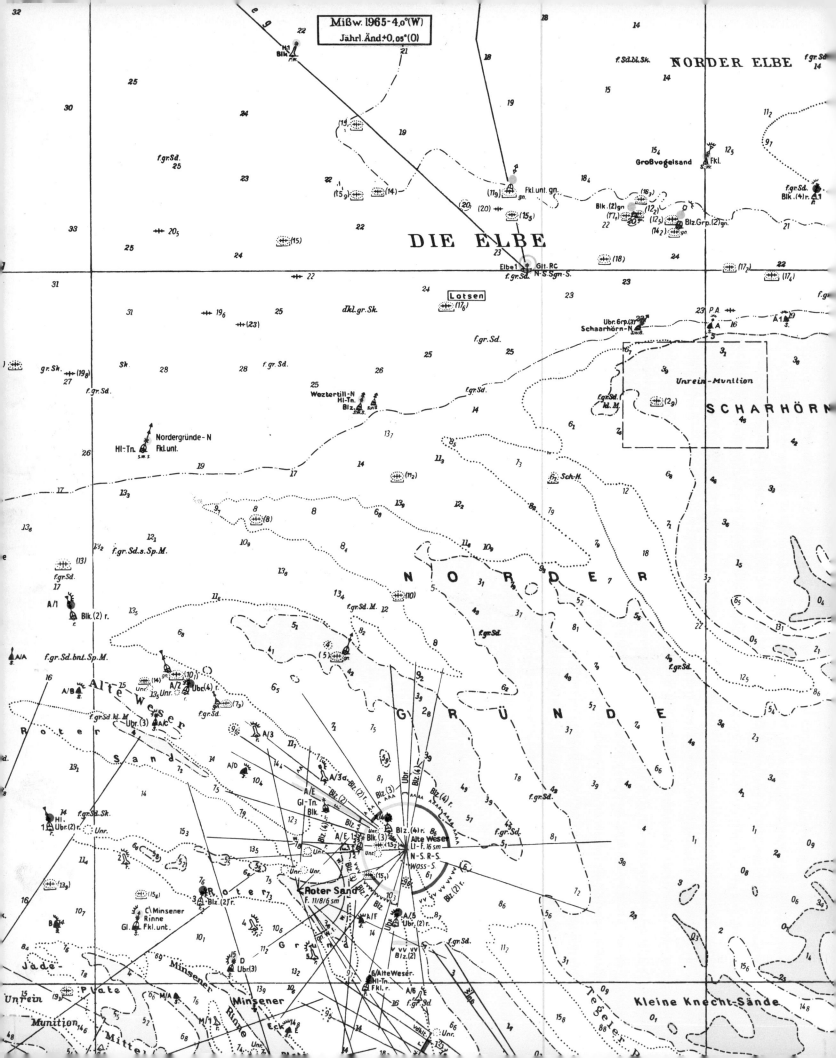

Sea chart of the Elbe estuary as used in modern shipping

on the flat charts, so on this one too, all longitudinal lines ran parallel to each other. To obtain the sought-for accuracy of angles, Mercator increased the distances of latitudinal measurements from the equator to the poles in proportion to the distances the meridians drew apart. This effective idea prevented what had hitherto been a stumbling block to navigation – there was no longer any distortion of depicted regions in higher latitudes. The chart met the requirement of accuracy and made a true plotting of courses possible.

Although it is true that Mercator's method deprived his chart of a unified scale, that did not diminish its usefulness and it has not been superseded since. To this day, the Mercator Projection is adhered to by seafarers throughout the world. Nowadays, up-to-date sea charts are published by hydrographic institutes on the basis of exact surveys and oceanic research. Maritime books as well as information bulletins about coastal changes, changes in fairways, beacons and other indicators complement these charts.

Compass-card with full and quarter divisions as used for centuries in shipping and can still be found on older vessels

Old Portuguese compass dating back to 1780
In the 10th or 11th century knowledge of a "North Indicator" came to the Arabs via the Orient and the Occident from China. This primitive instrument was improved on in the 13th century by the Italians who invented a ship's compass. In the Middle Ages the face of such compasses was often artistically decorated. The compass depicted here is adorned with the crown of the King of Portugal. Since in the Middle Ages the steering direction was not North, but East-West, that section of the compass face was more richly decorated.

COMPASS

On larger sailing ships in past centuries many captains had small compasses suspended in cardan joints from the bulkheads above their berths. That way they were always able to control their helmsmen, and course.

OCTANT

The octant is an instrument for measuring the angle of stars. It was invented in England in 1731 and, next to the compass, it became the most important aid in navigation, since it is mathematically possible to calculate the position of a ship by the angle of stars. Instruments preceding the octant were primitive in that they did not allow for accurate calculations, only the octant made such calculations adequate.

HOUR GLASS

In the past all ships carried two hour glasses. One of them always ran out within the half-hour at which time a ship's bell was sounded. At eight bells, or every four hours, a ship's watch was ended. The second hour glass was much smaller and ran out in 14 seconds. With it the ship's speed was measured, when the log was heaved overboard.

CELESTIAL GLOBE

With the aid of a celestial globe, which shows the entire sky on a sphere, the coordinates of stars can be precalculated which facilitates the locating of a particular star. This has proven helpful at times in thick weather when the clouds allow only the briefest observation of stars.

Whaler at the
Cape of Good Hope

SHARK ON THE HOOK

In the Caribbean there is a great prevalence of professional shark fishing. Cuban fishermen alone will catch up to 50,000 sharks a year. Almost nightly, the fishermen of the Cooperativa de Pescadores de Cojimar go out after these dangerous, vicious fish. Before the boats depart, hooks and lines are carefully prepared on which a bait of small fish is fastened. In every boat there are two men, and there are old and young men who pursue this adventurous catch.

Approximately five sea miles beyond the coast the lines are put out, left overnight, and hauled in at dawn. When they are hauled in on the hook, the dangerous and aggressive sharks are invariably clubbed to death. Eight to ten sharks is considered a good catch with which to return ashore. Usually they are fish of medium size, about two metres long, but I have seen an old fisherman who brought in a tiger shark of five metres length.

ERNEST HEMINGWAY
THE OLD MAN AND THE SEA

The old man made the sheet fast and jammed the tiller. Then he took up the oar with the knife lashed to it. He lifted it as lightly as he could because his hand rebelled at the pain. Then he opened and closed them on it lightly to loosen them. He closed them firmly so they would take the pain now and would not flinch and watched the sharks come. He could see their wide, flattened, shovel-pointed heads now and their white-tipped wide pectoral fins. They were hateful sharks, bad smelling, scavengers as well as killers, and when they were hungry they would bite at an oar or the rudder of a boat. It was these sharks that would cut the turtles' legs and flippers off when the turtles were asleep on the surface, and they would hit a man in the water, if they were hungry, even if the man had no smell of fish blood nor of fish slime on him.

"Ay," the old man said. "Galanos. Come on, galanos."

They came. But they did not come as the Mako had come. One turned and went out of sight under the skiff and the old man could feel the skiff shake as he jerked and pulled on the fish. The other watched the old man with his slitted yellow eyes and then came in fast with his half circle of jaws wide to hit the fish where he had already been bitten. The line showed clearly on the top of his brown head and back where the brain joined the spinal cord and the old man drove the knife on the oar into the juncture, withdrew it, and drove it in again into the shark's yellow cat-like eyes. The shark let go of the fish and slid down, swallowing what he had taken as he died.

Klaus H. Zürner: The old man and the sea

Several thousand seals still roam the North Sea coast. On fine days the seals use sandbanks above the water far out at sea to sun themselves. Seals are very shy, but also nosy. As they lie in droves, the bull, the leading seal, can smell approaching humans from far away, and then they abandon the sandbanks hurriedly and plunge back into the water. A solitary seal may be surprised and approached only when it sleeps.

DUTCH RESCUE SHIP ON THE WAY TO SUCCOR A SHIP

To rescue people in distress has always been a moral concept with inhabitants of the coasts. Plucky men have organized themselves in guilds and societies for the rescue of imperiled seamen. In the past, rescue crews had to take to open boats and row into the turbulent seas in order to help others. Nowadays almost unsinkable motor launches with the best of equipment lie in readiness along the most dangerous coastal areas, and even in the foulest weather these vessels are capable of assisting ships in distress.

North Sea fisherman

Today a diver will submerge in the water wearing little more than a mask and an oxygen bottle on his back. He barely notices the terrific water pressure on his body, and so he can move freely and without obstruction like the fauna of the sea. He discovers the wonders of the ocean: fabulous underwater gardens and edifices of corals, grottoes, caves and a colourful variety of sea life under water.

ACKNOWLEDGEMENTS

The following institutions and individuals, besides the author, supplied us with pictures and photos:

Cartographic Department of the Saxon State Library, Dresden 49, 52, 118/119, 139
British Museum, London 51
Picture Agency "Zentralbild", Berlin 46
Peter Steffen, Foreign Service Press Photos, Berlin 64
Press Agency "Novosti", Moscow 55, 57
Popper Agency, London 115
Tass, Moscow 114
Graphic Department State Museum, Berlin 166, 167
Institute of Egyptology, Humboldt University, Berlin 156
Blohm & Voss, Hamburg, HAPAG Works Photo 174
Flying Camera, New York 176
Andreas Feininger, New York 180/81
Cees van der Meulen, Hemstede 202
Jarrold & Sons, Ltd. London 61
Maurice Berney, Montreux 112/113
Sigi Köster, Munich 204
Hydrographical Service of the GDR 186/189/190
E. A. Seemann, Leipzig 116
Seebüll Foundation Ada and Emil Nolde, Seebüll 4/3
Seyler, Berlin, Drawings and nautical projections

ISBN 0 200 771689.1
LCCC No. 74–119606

Lay-out: Walter Schiller, Altenburg
Printers: Druckerei Fortschritt Erfurt

Copyright 1970 by Edition Leipzig
Printed in the German Democratic Republic

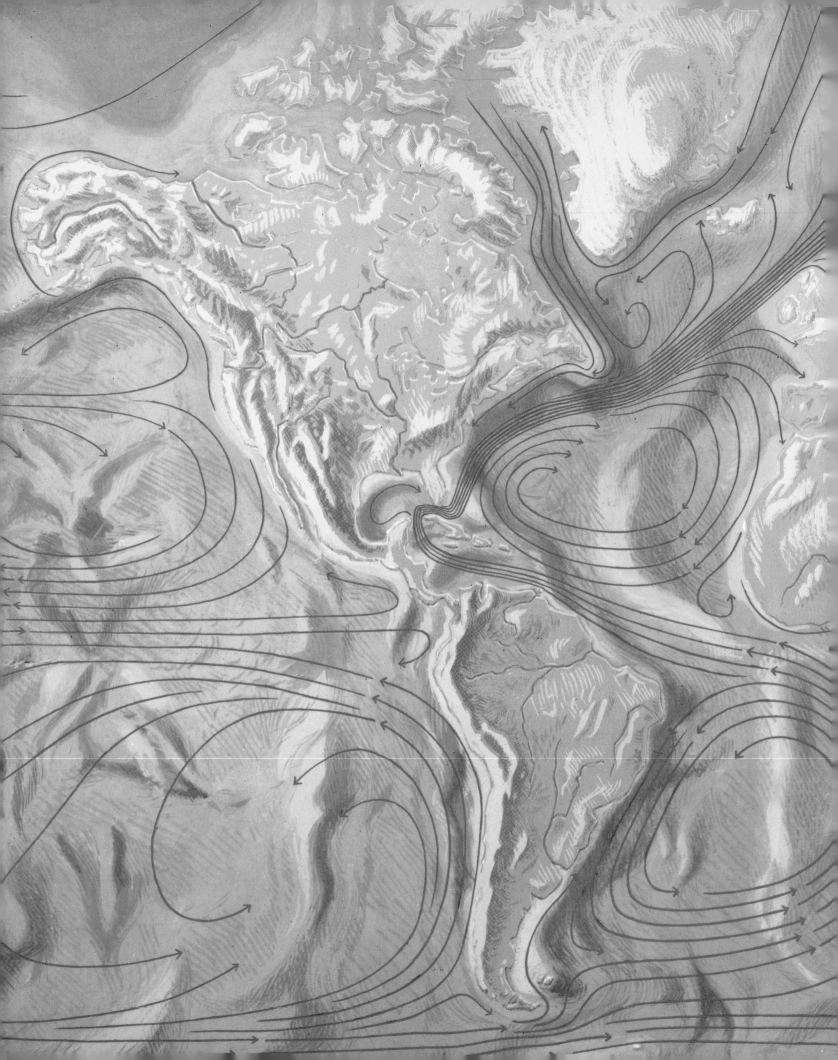